For Mia,
Aeran and Ander
~ Nic Jones

For Charlotte
~ Harriet Evans

LITTLE TIGER
An imprint of Little Tiger Press Limited
www.littletiger.co.uk
1 Coda Studios, 189 Munster Road,
London SW6 6AW
Imported into the EEA by
Penguin Random House Ireland,
Morrison Chambers, 32 Nassau Street,
Dublin D02 YH68
First published in Great Britain 2022
This edition published 2024
Text by Harriet Evans
Text copyright © Little Tiger Press Limited 2022
Illustrations copyright © Nic Jones 2022
A CIP catalogue record for this book
is available from the British Library
All rights reserved • Printed in China
ISBN: 978-1-83891-648-0
CPB/1800/2763/0524
1 2 3 4 5 6 7 8 9 10

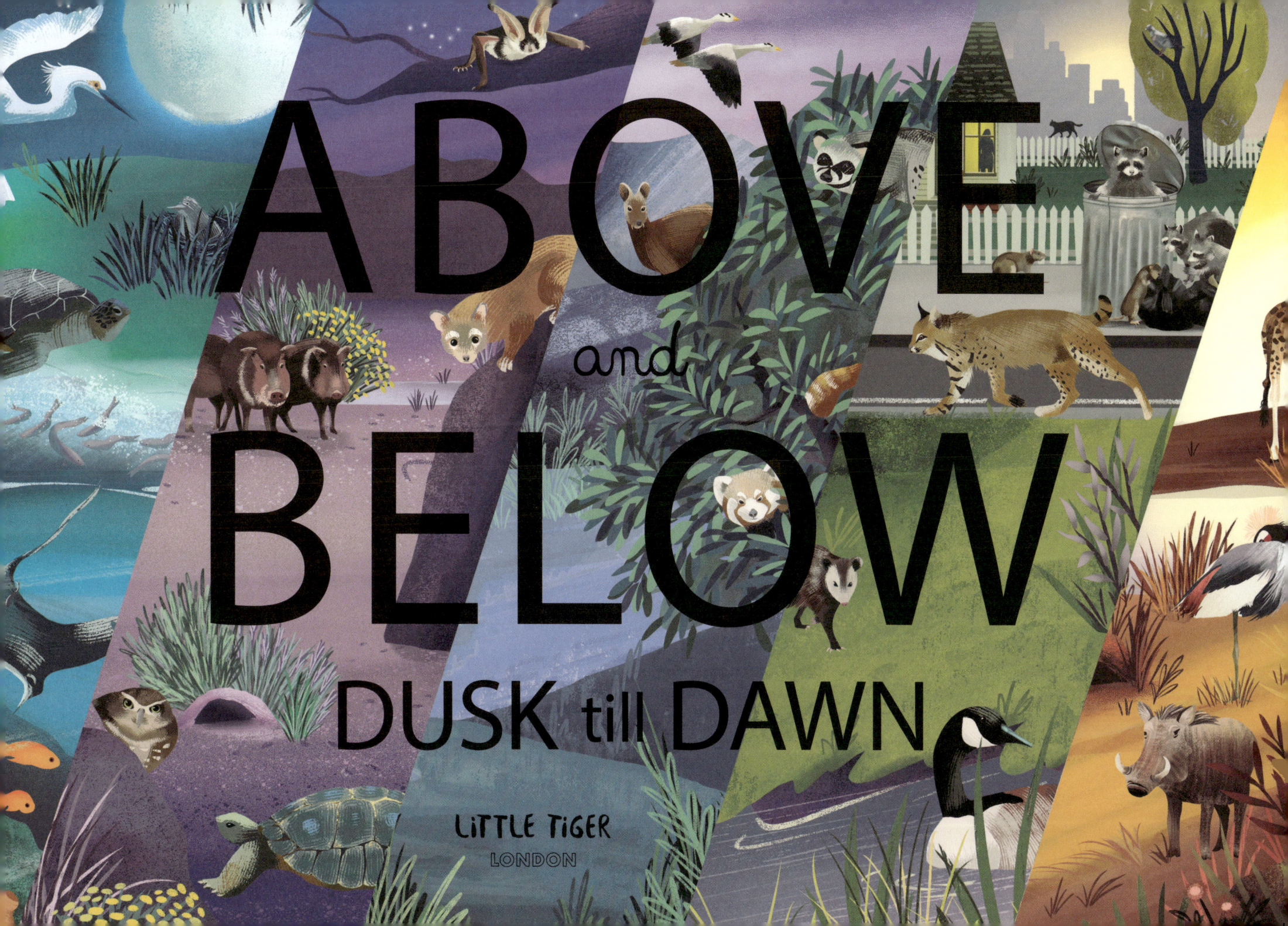

ABOVE and BELOW

DUSK till DAWN

LiTTLE TiGER

LONDON

Savannah

The golden hours of the savannah's sunset deliver respite from the day's heat. As some creatures bed down for the night, others are only just emerging from their cool burrows.

giraffe
To stay alert against predators, giraffes only sleep for minutes at a time and can do so standing up.

wildebeest
On the search for lush vegetation, more than 1.5 million wildebeest migrate around 1,000km (600mi) during the dry season.

grey crowned crane
As part of their mating rituals, grey crowned cranes dance, jump and bow to one another.

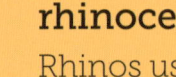

leopard
Notoriously good climbers, leopards carry their prey into trees so that ground-dwelling scavengers cannot steal their food.

bushbaby
A bushbaby looks after its ears when jumping between trees by folding them in close to its head.

bushpig
These creatures often follow monkeys, eating any fruit the primates drop when feeding.

rhinoceros
Rhinos use mud to protect their skin from bugs and the Sun's bright rays.

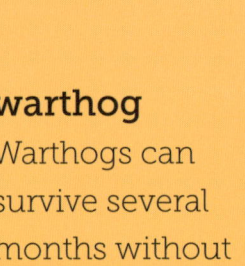

warthog
Warthogs can survive several months without water if need be.

European badger

Badgers are very careful to keep their underground homes – called setts – clean. They don't bring food in and only poo outside!

emperor moth

While female emperor moths only fly at night, males also venture out during the day.

honeysuckle

Honeysuckle gives off a stronger scent at night, attracting nocturnal pollinators, such as moths.

red fox

Despite hunting the same prey, foxes and badgers sometimes live together in badger setts.

European rabbit

Rabbits can turn their ears 270° to listen for predators.

roe deer

Unlike other types of deer, the male roe grows its antlers in winter, not summer.

Woods

As dusk descends, the woods come alive with scuffling and shuffling. From rabbits to roe deer, there are plenty of creatures hiding among the trees...

Eurasian beaver

Beavers construct dams in rivers to make calm pools where they can build lodges. The inside of each lodge is above water and sometimes beavers build a separate area by the entrance where they can dry off.

common toad

A toad can release poison from its skin to keep predators at bay.

grey squirrel

The main way squirrels communicate is through twitching their tails.

weasel

Weasels don't build their own burrows but take over the homes of animals that they eat. Their slim, agile bodies can slide into a variety of nooks and crannies.

hedgehog

By eating substances that are harmful to most creatures, such as toxic plants, hedgehogs create a poisonous foam to spread over their spines, possibly for protection.

great spotted woodpecker

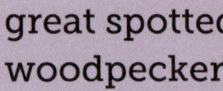

Woodpeckers have long tongues to catch insects hiding in trees.

hazel dormouse

It takes around 20 minutes for a hazel dormouse to nibble into the nut that is its namesake.

barn owl

Barn owls swallow small mammals whole and then throw up pellets of indigestible bone and fur.

Arctic tundra

Due to the Earth's tilt, the Arctic circle faces towards the Sun in summer and away from it in winter. This area has at least 24 hours when the Sun doesn't set around June and when it doesn't rise in about December.

northern lights

As charged particles from the Sun collide with the Earth's atmosphere, they create a cacophony of colour! This is called the northern lights or aurora borealis.

snowy owl

To protect themselves from the cold, snowy owls have plenty of feathers, even on their feet!

musk ox

Both males and females have horns that never stop growing.

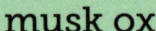

Arctic fox

Arctic foxes eat the remains of polar bears' meals when they cannot find food on their own.

reindeer

A reindeer's eyes change from gold to blue in winter to reflect less light, helping them see better during the darker months.

snowshoe hare

Large back paws help snowshoe hares bound easily over snow.

Arctic wolf

Arctic wolves have two layers of fur to stay warm and dry.

Arctic tundra

Due to the Earth's tilt, the Arctic circle faces towards the Sun in summer and away from it in winter. This area has at least 24 hours when the Sun doesn't set around June and when it doesn't rise in about December.

northern lights

As charged particles from the Sun collide with the Earth's atmosphere, they create a cacophony of colour! This is called the northern lights or aurora borealis.

snowy owl

To protect themselves from the cold, snowy owls have plenty of feathers, even on their feet!

musk ox

Both males and females have horns that never stop growing.

Arctic fox

Arctic foxes eat the remains of polar bears' meals when they cannot find food on their own.

reindeer

A reindeer's eyes change from gold to blue in winter to reflect less light, helping them see better during the darker months.

snowshoe hare

Large back paws help snowshoe hares bound easily over snow.

Arctic wolf

Arctic wolves have two layers of fur to stay warm and dry.

Jungle

The jungle is the backdrop for some of nature's most ferocious hunts. Smaller animals have adapted to avoid their attackers by venturing out only at night. But, just as these creatures have changed their hours, predators now lurk in the darkness, ready to pounce...

potoo
The potoo bird has tiny slits in its eyelids that allow it to see even with its eyes closed.

spider monkey
Spider monkeys are very social – they even hug each other! They live in groups of around 35 individuals, though they will split off into smaller sets to find food and eat.

vampire bat
Vampire bats are surprisingly altruistic. They will give some of their own food to companions who have had a less successful hunt and will occasionally look after orphaned young.

firefly
Fireflies are thought to flash as a way to attract a mate or communicate. Different species have different flashing patterns.

anaconda
An anaconda can hold its breath underwater for around ten minutes.

margay
Adept climbers, margays can hang from tree branches using only one paw.

tapir
When under threat, tapirs escape to the water as they are fast swimmers. They sometimes use their long noses like snorkels.

fungus
There are over 70 known species of fungus that glow in the dark.

piranha

Despite their blood-crazed reputation, piranhas are orderly eaters, taking turns feeding on their prey.

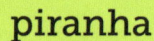

jaguar

The jaguar is the cat with the most powerful bite – helping it take down animals three or four times its own weight.

river dolphin

Though already pink, the Amazon river dolphin blushes a more intense shade when it feels surprise or excitement.

giant river otter

Living up to their name, giant river otters can grow to 1.8m (6ft) and eat as much as 4kg (9lb) of food a day.

capybara

Weighing up to 66kg (146lb), capybaras are the world's largest rodents.

caiman

A female caiman can lay 65 eggs at a time.

Sea

During the day, tiny plants called phytoplankton soak in the Sun on the surface of the sea. At night, their animal counterparts, zooplankton, rise up to feed on them. This movement triggers a mass migration from the depths as the whole food chain is reeled upwards.

night heron
Night herons often nest with other bird species and even raise the chicks of these different birds on occasion.

sperm whale
Sperm whales likely need the least sleep of any mammal – spending 93% of their time awake.

grunion
These small fish leave the water en masse to lay their eggs after a full or new moon when the tide is at its highest.

northern elephant seal
Northern elephant seals migrate between California, where they breed and moult, and Alaska, the location of their feeding grounds.

turtle
Turtles always lay their eggs on the same beach where they hatched, even if they have to travel thousands of miles to do so.

dinoflagellate
Dinoflagellates are very small aquatic organisms – they are only a single cell in size. Some glow at night, causing the sea to sparkle.

snowy egret
Though egrets are usually awake during the day, they come out at night to hunt when grunions are laying their eggs!

manta ray

Manta rays regularly visit the coral reef so that the wrasse fish living there will remove any parasites or dead skin from them.

great white shark

The largest predatory fish, a great white shark can sense a single drop of blood in 100l of water.

midshipman fish

Male midshipman fish try to attract females by humming at night. They build nests near the shore to protect any eggs.

brittle star

Some brittle stars glow in the dark. If they're under threat, they detach an arm, which will flash and distract predators while they escape.

moray eel

The moray eel has double jaws; one set to trap its meal in its mouth and another to drag it into its throat.

coral

During the night, coral extends its tiny tentacles to feed.

Desert

With the sweltering Sun above and scorching sands below, the desert is inhospitable to most animals during the day. But once night falls, along with the temperature, these arid areas teem with life.

queen of the night
This cactus has white flowers that only bloom for a single night.

spotted bat
To navigate around objects in the dark, bats emit high-pitched noises and wait for the sound to return to them. This process is called echolocation. Unlike most bat species, the spotted bat's noises for echolocation are audible to humans.

Mojave rattlesnake
As the name suggests, this snake has a 'rattle' on the end of its tail to deter predators.

bark scorpion
A female scorpion can give birth to up to 35 offspring. They ride on her back until they are a couple of weeks old.

Gila monster
One of only a few species of venomous lizards, the Gila monster is a slow predator so it uses its venom to immobilise prey.

collared peccary
A pig-like animal, the collared peccary has a scent gland on its back. It uses this to mark its territory and others in its herd.

ringtail
Always on the move, a ringtail is unlikely to spend longer than four days in any den.

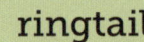

bar-headed goose

Soaring at altitudes over 7,000m (23,000ft), bar-headed geese have the highest migration flight of any bird.

Himalayan palm civet

Like skunks, Himalayan palm civets can produce a foul-smelling spray to warn off predators.

musk deer

Instead of antlers, musk deer have long fangs protruding from the mouth. Males use these when vying with each other for a female.

yak

Yaks can withstand temperatures as low as -40°C (-40°F) and will even swim in partially frozen water.

Saussurea

This plant's woolly covering may help it survive the freezing night-time temperatures.

Himalayan Mountains

A towering mountain range, the Himalayas boast the highest peaks in the world. The vertiginous landscape may seem unwelcoming but it cradles some of the world's rarest and most wonderful creatures.

pygmy hog

These hogs range in size from 55-70cm (22-28in), making them the smallest pigs in the world.

Malayan porcupine

As well as looking formidable, the Malayan porcupine's spines create a rattling noise when shaken to deter predators.

golden langur

Golden langurs are one of the most endangered species of primates.

red panda
Red pandas can turn their ankles 180° for better grip when moving down trees.

Himalayan swiftlet
The Himalayan swiftlet is one of the few bird species that uses echolocation. It helps the bird navigate the cave systems where it roosts.

snow leopard
Snow leopards have very thick tails to keep them warm, help them balance and to store fat for when food is short.

Himalayan wolf
Due to its high-altitude habitat, the Himalayan wolf has adapted to use oxygen more efficiently than other grey wolves.

Asiatic black bear
In the winter, most Asiatic black bears hibernate in caves or tree holes. During this time they don't eat, move or even poo.

prairie hare

Parents leave their young dotted around the city away from their home to make them harder for predators to track.

coyote

Though coyotes in rural areas are typically awake during the day, the ones that have colonised cities tend to be nocturnal so that they can avoid humans.

skunk

Before shooting its stinky spray, a skunk slaps its tail on the ground and stamps as an intimidation tactic. The western spotted skunk will even do a handstand to show off its markings and to appear larger.

screech owl

When in danger, a screech owl camouflages itself among the trees. It may make itself appear thinner, close its eyes and even sway like a branch in the wind.

City

As cities sprawl outwards, more and more animals creep in, attracted by the readily available source of food: human rubbish. The constant bright lights and a desire to avoid humans have changed many animals' natural sleeping patterns in urban environments so that they become citizens of the night.

opossum

If opossums feel threatened, they play dead and even give off a smell similar to a dead animal.

Canada goose

Geese tend to sleep on water, taking turns to watch over the flock.